Bibliographic information published by the German National Library:

The German National Library lists this publication in the National Bibliography; detailed bibliographic data are available on the Internet at http://dnb.dnb.de .

Imprint:

Print and binding: Books on Demand GmbH, Norderstedt Germany
ISBN: 9783346220233

This book at GRIN:

https://www.grin.com/document/591052

Philipp Straßer, Eva Missoni

GIS Concepts For School Children. Creating A Scavenger Hunt For Children With ArcGIS Collector And Survey123

GRIN Verlag

Introduction of GIS concepts to school children

- Creating a scavenger hunt for children with ArcGIS collector and Survey123

Created by: Eva Missoni and Philipp Strasser

Content

Table of figures

Abstract

A scavenger hunt can be a ludic and active approach to address pupils and transport educational and subject content. Thus, the scavenger hunt we designed aims to familiarize pupils of the age class 10-12 years with GIS and the possibilities GIS offer in a playful manner. It includes 5 stations which are addressed by following an instruction-sheet with QR-codes leading to the respective places and posing questions and tasks.

For the scavenger hunt there are two datasets that have to be handled: The tasks for each station including the submitted answers and the tracks of the groups. The Results and the tracks should also be made available to the Z_GIS geoportal following the guidelines of open standards and interoperability. To fulfil these requirements a combination of two ESRI software products was used: ArcGIS Collector and Survey123. The tracks can be accessed as web feature service via REST URL whereas the survey answers are directly integrated into a dashboard with a map. So, the results are immediately updated and available in ArcGIS Online for inspection, comparison and discussion.

1. Introduction

The initial task in the lecture SDI Services Implementation was to develop an own project including an SDI-strategy and standardized geoservices for the area of Radstadt. The created services shall be prepared appropriately to communicate to school children what GIS is and are feasible to be done at the GIS-day in May 2019.
The scavenger hunt we created for this event aims to provide a playful introduction to GIS-content for pupils. Therefore, we also encourage pupils to actually use "real" GIS-software. As they are about 10-12 years old, we decided to reduce complexity of the software as much as possible and adding an appealing, child-oriented storyline while still having a serious GIS-environment to conduct the scavenger hunt.

Some words about the Software used:

We decided on a mixture of two ESRI-apps as basis for the scavenger hunt: ArcGIS Collector for taking the tracks of pupils walking and Survey123 for fulfilling tasks and submitting answers. Each app has a clear function within our project and can be applied simultaneously, as ArcGISCollector will work in the background.

2. Project idea, content and scope

To offer a project appropriate to the age of school kids between 10-12 years, a kind of scavenger hunt was developed to communicate what GIS is and can do. The scavenger hunt is designed to take place in small groups of 3-5 kids, which are equipped with tablets and an instruction-sheet.

As one presentation-session is intended to take about 45 minutes, the scavenger hunt is designed to take place on the school compound of Castle Tandalier. Thus, long distances to walk can be avoided as well as for liability reasons.

A scavenger hunt provides the possibility to transport information and aims of GIS while at the same time it is a ludic approach to introduce to GIS: As own action and physical activity is required, the project is well suited to be performed between more theoretical items on the agenda of the GIS-day. The ludic approach is underpinned by the introduction of the ghost GISO, who leads through the tasks in order to have a red thread and storyline for the scavenger hunt. It was our concern to offer an approach which is perceived as fun, but at the same time transports serious messages, such as the usage of different technologies to display fields of GIS. Tasks and questions at different stops are designed easy to handle but vary in data types obtained by the school children to have diversity in the results that are to be discussed in short after the practical scavenger hunt. The scavenger hunt tasks are based on ArcGIS software to display the results and collect data deriving from the tasks: Once activated, the ArcGIS Collector application collects the tracks while the kids are doing the scavenger hunt tasks, which are provided as Survey123-forms, accessible via QR-codes provided on the instruction-sheet. The collected tracks and data from the tasks are directly integrated and displayed in an ArcGIS Online WebApp as visual basis for discussion.

3. Methodological approach

a. Workflow and SDI-Approach

For the scavenger hunt there are two datasets that have to be handled: One the one hand the tasks for each station including the submitted answers and on the other hand the tracks of the groups walking from task to task. The used software should be easily compatible with the geoportal of ZGIS which is an ESRI portal that is locally maintained by the University of Salzburg. The tracks as well as the tasks and their results should finally be stored at the geoportal to be open accessible within the organization in order to ensure the reusability and adaption for other school groups than the school of Castle Tandalier in Radstadt.

Since the tracks and the tasks impose different requirements to the software, we decided to use two different types of Applications for data acquisition within the ESRI universe: ArcGIS Collector for the tracks and Survey123 for the stations. The next chapters deal with these two apps in more detailed way. The idea is to run both apps at the same time, the Survey123 app as main interface for the scavenger hunt in the foreground and the collector app recording the tracks in the background. The tracks should be stored as feature service in the ArcGIS Server of the geoportal and be public available as Web feature Service. They are accessed via Rest URL into the Web map with the results as hosted feature layer. The Survey123 task results cannot be published as web feature service because they are directly loaded into automatically created folders in ArcGIS Online. Finally, the feature layer can be loaded into a web map application or a dashboard to bring both outcomes together and present it nicely. Figure 1 summarizes the approach that was used in this case.

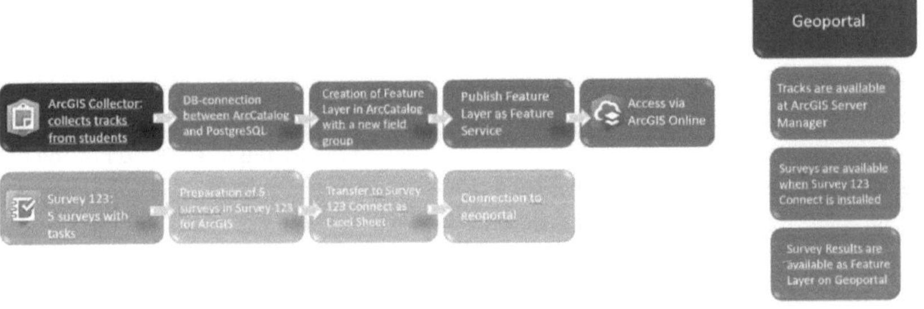

Figure 1 Scavenger Hunt SDI approach depicted in a workflow (source: own illustration)

b. Comparison of ArcGIS Collector and Survey123

ArcGIS Collector and Survey123 were both used for this project due to the fact that they have some strengths and weaknesses that make each of them better for a certain purpose. Used together they can compensate the drawbacks of the other app and were in our case a useful to fulfil the requirements of this project. However, they also have some things that they have in common:

- Collector and Survey123 are both applications for mobile data collection
- They are ESRI products which makes them easier to use together with other ESRI software
- An ArcGIS online Account or Portal is needed to store/view the results
- Data collection is also possible offline, the data is uploaded later when connection to the internet is established again
- Editor tracking is possible
- Both apps allow to add photos
- Both allow capturing of new data as well as editing of existing data

The most obvious difference between the Apps that initially catches one's eyes is the layout of the data acquisition: While ArcGIS Collector is map centric, the Survey123 App offers a form centric interface. That means in Collector you always see your input/tracks situated on a map and you even need a ready web map in ArcGIS Online to start with data collection. In Survey123 you start with styling and designing the form of the surveys and tasks, the acquisition is also done within this layout and only afterwards you can look at your georeferenced results on a map. Survey123 only displays the results as survey points whereas with collector one is able to acquire also line features and polygon features. For ArcGIS Collector you need an ArcGIS account to collect data, in Survey123 there is at least the possibility to take part of a survey anonymously in the web version of Survey123.

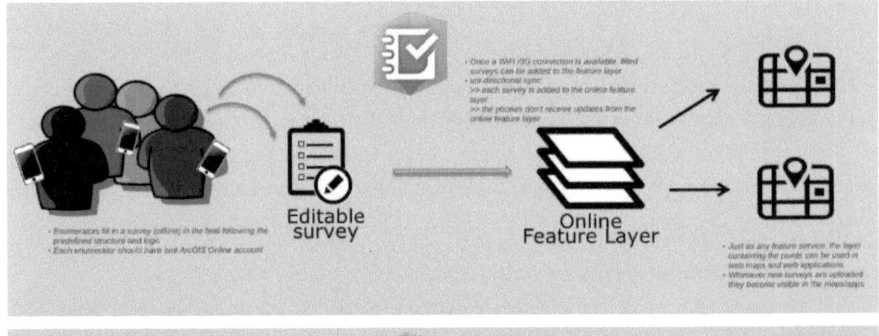

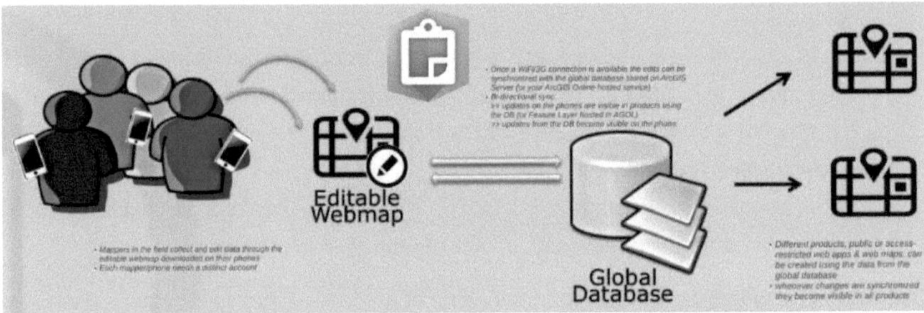

Figure 2 Comparison from Collection to Publication (Source: Alschner 2016)

The big advantage of Survey123 is to customize your layout of the surveys, allows skipping questions, determining required questions, applying expressions and rules and labelling of the questions in different languages. As described in the previous chapter they are also different in dealing with the collected data (Figure 2). ArcGIS Collector is more open and flexible to the storage and access to the data on a Geo Server. The App even allows a two-way communication and updating of the data.[1]

To sum up, Survey123 was the first choice for the tasks that are given to the pupils since the interface is highly customizable. It allows a variety of survey question forms like multiple choice, photo input or setting of a marker on a map. ArcGIS collector offers the possibility to map the tracks as line features and is better in publishing the data.

c. Prepare acquisition of tracks with ArcGIS Collector

Before you can start with collecting data with the ArcGIS Collector you first have to make a feature layer where the data is inserted and a web map that displays the feature layer. For this project we wanted to make the feature layer public available according to open geo standards. To accomplish these requirements, a PostgreSQL connection was established with

[1] Alschner,S. (2016): Test: Survey123 and Collector for ArcGIS. Retrieved from: http://blog.cartong.org/2016/02/18/survey123-collector-arcgis/. Accessed: 10.02.2019

the ZGIS database within ArcCatalog. The next step involves the creation of a new feature layer, in this case a line feature as the tracks should be displayed as lines. Additionally, a field for the group name of the children was added to make sure that we can differentiate their tracks afterwards. ArcGIS Collector should ask for the group name before starting with acquisition. The feature layer which is now empty is then published to the ArcGIS Server as Feature Service. Here you can define that it should be available as feature service and it is also possible to enter metadata. By logging in into the ArcGIS Server of ZGIS the REST URL can be retrieved to access the data as web feature service. The feature layer was then added to ArcGIS Online and also into the final web app as hosted feature layer. By logging in with the same account in ArcGIS Collector it should be available as target for inputting tracks.

Figure 3 Tracks on ArcGIS Server Manager (source: ArcGIS Server Manager, © ESRI)

d. Decision between Web and Connect version of Survey123

ESRI actually provides two versions of Survey123 to work with:

- There is the web version with predefined question types of which a survey can be composed of. The creation of the survey as well as the editing, storing and sharing is all done in a Web Browser combined with an ArcGIS Account.
- Survey123 Connect is the locally stored counterpart of the Survey123 web version, which is based on Excel Sheets. Survey123 Connect goes hand in hand with the mobile Survey123 application download and fill out surveys from Connect but also from the web version. Survey123 Connect offers the possibility to prepare more advanced question types than the given ones in the web version and it also allows to apply complex rules and more depth in customization.

For the scavenger hunt we went with the web version for several reasons: The web version is the more intuitive, user-friendly one and does not require much foreknowledge in comparison.

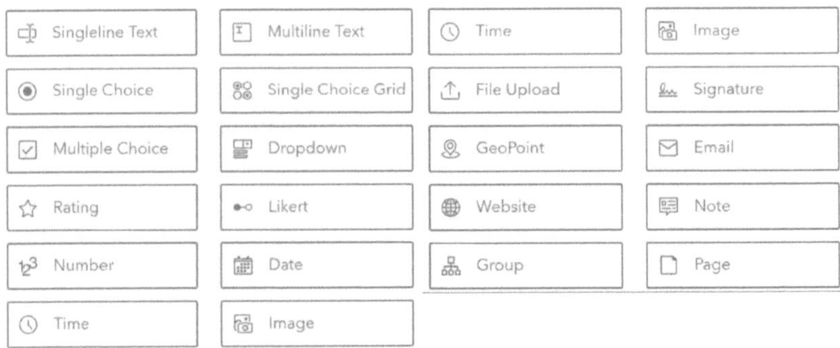

Figure 4 Predefined elements of the web version (source: Survey123 © ESRI)

Furthermore, the provided elements offer a big variety and were more than sufficient for our purpose (Figure 4).

The default interface of the web version is also more adapted to the usage with mobile phones and tablet which are our target devices. The web version allows also to preview the survey for tablets, mobile phones and computers and therefore it is easier to adapt the layout.

There are even some things that are only possible in the web version: For instance, generating QR codes directly, which we definitely wanted to use for opening the tasks, as according our experience the QR-codes run more stable than providing URLs. Another issue is that you can open the surveys within a web browser in contrast to the connect version where you need an ArcGIS Account and the mobile app to fill out the questions. This is a major advantage because no installation of the app is needed, and pupils could even open the app on their private mobile phones. One is also able to design an own "Thank you for your submission" screen which is also a nice extra that is only provided in the web version. Finally, editing in Connect has to be done in the Excel sheet and then be applied to the survey to see them whereas in the web version the changes can be viewed directly and interactively in the editor tab.

4. Generation of tasks in Survey123

The generation of the survey involved the finding of suitable tasks that can be achieved within the grounds of the school at Castle Tandalier and within a certain amount of time. We tried to use a variety of the predefined elements to keep the tasks interesting. There are five tasks: one for each cardinal direction and one task in the centre at the castle. Each task has a unique colour to be easily recognizable (e.g. North → blue).

The structure of the surveys is pretty much the same as shown in figure 5: Header with the unique colour and the name of the survey followed by a task description by the ghost in German. Each survey also has an image that illustrates the task and is also used as thumbnail of the survey.

The next element to insert is the **name of the group**. This is necessary as there are survey points from five tasks and up to five groups, otherwise it would be impossible **to distinguish them** in the final web map. The field properties are therefore set to required. The following tasks were formulated:

- North: Starting and end time of a walk with 200 steps
- East: Putting marker on a ping pong table on an orthophoto of the area around the castle
- South: Identify mountain tops in sight distance according to the weather
- West: Estimating the size of a tennis court
- Castle: Take a photo of an object that is older than 50 years

The initial submit button was further modified to a German version. After the data is sent a customized "Thank you screen" (figure 5) appears on the screen. After the generation and publication of a survey, Survey123 automatically creates a folder for the survey results and a feature layer in ArcGIS online. The web version provides multiple options of sharing the surveys: as Link that opens either the survey in the mobile app or in the web browser, as QR code or embedded in a website. In our case we chose the QR Codes which were designed with the QR Code Generator (http://goqr.me/) in the according colour.

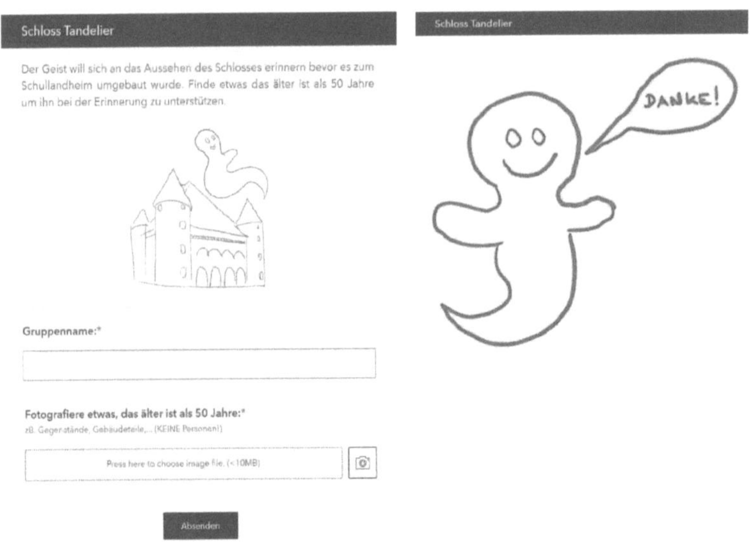

Figure 5 Survey Layout and Submission screen (source: own illustration)

5. Integration and presentation of the results

To establish connection between the Collector app and the tracks feature layer a web map is already needed. Into this web map also the several feature layers of the survey, that are still empty, were loaded. As a base map the decision was made to use a satellite image as this corresponds more to the perception of the children. Another issue was the scope of the results, the scavenger hunt is in a very small area around the Castle Tandalier. A vector-based map like a land use map or a map with the buildings would not make sense in this case as you would only see grassland and the building of Castle Tandalier. So, it would be impossible to see the ping pong table, which you should find in task "Osten". The normal imagery base map of ArcGIS offers a lot more details but even that is to coarse regarding the area of interest. Therefore, we decided to integrate the orthophoto of Land Salzburg which is provided as web map service by our colleague Simon Ecke for the municipality of Radstadt and Altenmarkt. The extent was set to the area around the castle, so the tracks and the survey results. Will initially be shown.

To present the results we also wanted to show a few of the task results also. In a web map application, you could click on the survey points or open the attribute table to view the answers. Another possibility would be the analyse and data page in the survey123 web version, which shows the number of participations as well as the sent answers already packed into diagrams and statistics. The best option in our opinion was a combination of the previously stated possibilities: a map view of the results and tracks that is clickable to get further information about the answers and some of the questions shown in diagrams. This was realized with the Operations Dashboard which is also a software provided by ESRI. The header was designed with the ghost thumbnail of Castle Tandalier and the title was changed to German. In the left diagram boxes the outcomes of the mountain tops should be shown as a pie chart and the size of the tennis court as a line chart. In our case it is empty, there is only the result of our test run in it to illustrate the idea. On the right side is the map which should show the tracks and the survey points, which should be also coloured after group names to be better distinguishable. It is also possible to hide or add layers and the points and tracks as well as the map itself is interactive.

Figure 6 Dashboard of Results (source: ArcGIS Dashboard, © ESRI; basemap: Land Salzburg, DigitalGlobe, Microsoft)

6. Dissemination and pedagogical approach

Subsequently, the procedure of the scavenger hunt in Radstadt is described in detail, including lessons learned by pupils.

Preparation:

- 5 tablets with ArcGIS Collector App and Survey123 → Login to an ArcGIS Online account.
- Spray/mark the compass points in front of the Castle Tandalier, so that pupils have orientation in which direction to walk conducting the tasks.

Procedure of conducting the scavenger hunt, designed for an estimated timeframe of 45 minutes:

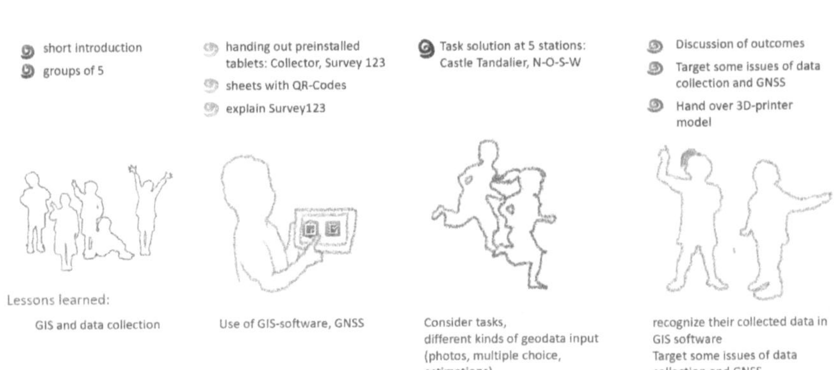

Figure 7 Pedagogical approach and lessons learned (source: own illustration)

Introduction (5 minutes):

- Initially we will provide a short introduction on what the scavenger hunt is, how it will be conducted and what data collection is and means for GIS.
- Then we ask the class to split in groups of 3-5 pupils and think of a nice of funny (and no offensive) names for their teams.

Lessons learned: *pupils get to know about how GIS and data collection are connected and how it can be conducted*

Instruction (5 minutes):

- 5 tablets of Z_GIS will be handed out to the teams (its good to have 1-2 in reserve, as one tablet often has "issues" and does not work properly). Check that there is about an equal distribution between girls and boys using the tablets. The tablets have ArcGIS Collector and Survey123 preinstalled.
- Also, paper sheets with coloured QR-Codes are handed out. These QR-Codes refer to the 5 stations of the scavenger hunt respectively: The Castle as Centre, North, East, South, West
- A short introduction of to handle Survey123 with the QR-Codes and how to fill the form.

Lessons learned: *introduction to use of GIS-software and elimination of inhibitions in regard to the use of GIS-technique*

Practical part (20-25 minutes):

- instruct pupils to open ArcGIS Collector App and the "Results Scavenger Hunt" map, enter their group name and enable STREAM to collect their tracks.
- Then teams go out, spread to all directions and solve the tasks at each station respectively according to their provided QR-Code-Sheet.

Lessons learned: *consider how to solve the tasks, get familiar with different kinds of geo-data which deal as inputs and answers to the tasks and questions, such as photos, multiple-choice and single-choice questions, estimations*

Discussion (10-15 minutes):

- When groups return, ArcGIS Collector tracks are synchronized: Tip on "Synchronize (x)" in the overview, so the data gets uploaded to ArcGIS Online immediately.
- The Survey123 data is updated immediately when filling out the form and pressing the "Submit"-button.
- We start a discussion round to take a look at the different data types collected, issues that might arise doing data collection and using GNSS for tracking
- Finally, pupils get handed over a 3D-Modell of Castle Tandalier as a giveaway and memory for the GIS-day taking place in Radstadt

Lessons learned: *recognize and compare their collected data in GIS-Software, address issues of data collection and accuracy of GNSS*

A first print of the 3D-Model was already conducted:

Figure 8 3d Model of Castle Tandalier as giveaways for the classes. (Realisation and photo credits: Peter Jeremias)

7. Testing of the scavenger hunt

Testing of the scavenger hunt did not take place in Radstadt at Castle Tandalier but was performed in the environment of the University of Salzburg, therefore we picked fictional places to fulfil the tasks. In order to test how the scavenger hunt works on different platforms, we conducted the testing with a tablet and a smartphone. It turned out that both devices worked out well, with the only difference that the tablet provided more exact tracking of our walk.

In the beginning we enabled line tracking in the ArcGIS Collector App by opening the "Results Scavenger Hunt" map, adding a new line layer with (1), entering a group name for later identification of who took which route and pressing the STREAM Button (2).

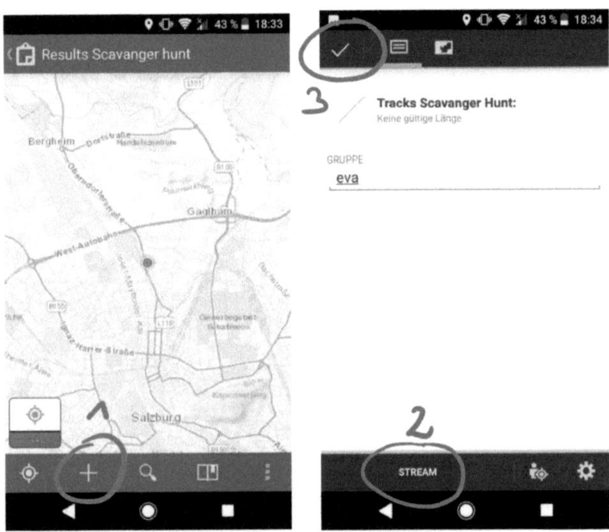

Figure 9 Streaming of tracks in ArcGIS Collector (source: ArcGISCollector, © ESRI; basemap: ESRI)

The app then did not need further attention until the end of our hunt (3) to save the line-feature, as it operates in the background and Surveys can be done at the same time.

Then, according to the instruction on the sheet, the first QR-code sent us to find, photograph and submit something older than 50 years (compare: Figure 5).

The uploaded image then is visible in the WebApp by clicking on the following link:

Figure 10 Vizualisation of collected point data in ArcGIS Online (source: ArcGISCollector, © ESRI; basemap: Land Salzburg)

Obvious in this map is also, that the red and yellow tracks show different accuracy of GNSS tracking our walks. E.g. the passage through the 2 buildings is quite accurately depicted by the red line, while the yellow one "crosses" a building. Such issues can be addressed in the discussion after conducting the scavenger hunt in Radstadt.

Figure 11 Complete Test Walk for the scavenger hunt (source: ArcGISCollector, © ESRI; basemap: Land Salzburg)

Whereas in other areas, the yellow line is more accurate, especially along the Alterbach, where the red line (mobile phone streaming of position) seemed to stream location data and connect it with lines in a quite greater interval than the tracking with the tablet (yellow), although highest accuracy was enabled for location tracking in the mobile phone settings:

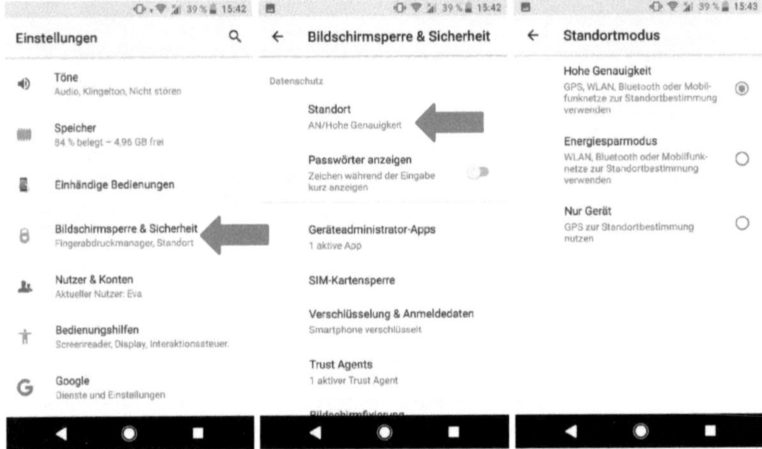

Figure 12 Mobile Phone settings of an Android phone to set accurracy of positioning (source: own illustration)

Proceeding with the scavenger hunt tasks we realized that we had to alter certain minor aspects of how we suggested the tasks: e.g. when the ghost asks to walk 200 meters and take the time for this, we changed it to 200 steps, as it won't be clear to the pupil how far 200 meters is when walking outside without additional markings.

Conducting the scavenger hunt was helpful to check which aspects had to be refined and which worked out pretty well. Generally spoken, the conduction went very smooth apart from some georeferenced points, which are saved correctly but displayed wrong in the map, while most were placed to the correct location. After the trial, we decided to implement a field for group names also in the surveys.

When returning to the PC, visualization of the different points as well as tracks can be distinguished via the group names. In addition, we changed the displaying of results from a conventional WebApp to a Dashboard, where also results can be displayed accordingly when they are uploaded in Radstadt.

8. Problems, challenges and solution management

As with any project, several challenges and issues arose. These can be distinguished between limitations in the functionality of the software, issues that we could solve or found a workaround for and challenges with external factors and technical obstacles (e.g. server-down-times and non-availability of services).

a. Limitations in the functionality of the Software

Initially we were thinking of only using one software as it should be as easy as possible to handle for the school children at the end. There were three different applications that came into question: Survey123, ArcGIS Collector and Trek2There which all are provided by ESRI. So, we had to test them regarding functionality in both ways, they should be easily usable for schoolchildren and offer a nice interface for them but also manageable for us to deal with them to achieve the creation of a scavenger hunt. Survey123 even consists of two different versions which had to be compared according to their pros and cons. Trek2There was excluded fast as it offers only a compass guiding to coordinates which was too limited in terms of a good interface for the tasks. Survey123 could serve for the purpose of presenting a customized interface but it was not feasible when it comes to mapping the tracks. As Collector and Survey123 can run concurrently on a tablet or a mobile phone we finally agreed on using both to combine their strengths.

b. Problems with Integration of Surveys to the geoportal and transferability

First, as we wanted to get familiar with Survey123 we tried the software and its functionalities on our personal ArcGIS Accounts. The surveys were all designed, finalized and tested in this environment. To make them available for the organization and also accessible for the other course participants our final step was to transfer the surveys with the tasks to the Survey123 of the ZGIS geoportal. It is also important because the folders of the results are automatically created in the linked ArcGIS online Account and therefore in our personal account. They should be integrated on the geoportal as well to be combined with the tracks.

Figure 13 Messed up layout (source: own illustration)

At the Survey123 Website there is no possibility to switch to a portal, the log in screen only offers a log in with normal ArcGIS online account, ArcGIS enterprise, Facebook and google Accounts. Interestingly, it seems that the geoportal has access to the Connect version of survey123 but not to the web version, which made the transfer of the surveys much more complicated. As solution we tried to move the surveys to the Connect version. Survey123 Web is capable of many options to share the survey but not the editable survey form. In addition, the save as button only copies the survey, so we had to download them to our private account in Survey123 Connect. A folder with all the information is created locally on the desktop, containing an excel sheet that shows the structure of the survey in a table, a media folder with the images and some other data.

Unfortunately, information about the layout gets lost in the transformation: The colour of each survey was set back to the default dark green, the font size got way too small to be readable on a mobile phone screen and the photos were tripled and maximized to the whole extent. (Figure 13) Moreover, the submit button and the submit screen which was also customized were gone. To cope with this issue the Excel sheet was first transferred to Geoportal Survey 123 Connect. We feared that further adjustments would also get lost within this step, therefore the design should be fixed in the final storage location. In the excel sheet the two extra entries for pictures (Figure 14) were deleted, as well as information added into the HTML code of the font. The colour could be changed in Survey123 Connect directly in the style page. Unlucky, Survey123 Connect does not provide the option to generate QR Codes directly. Only a Link to open the survey in the app can be converted into an QR code by an open source QR code generator. The other option would be to download the survey directly into the app. Opening a survey in the web browser was not manageable at all.

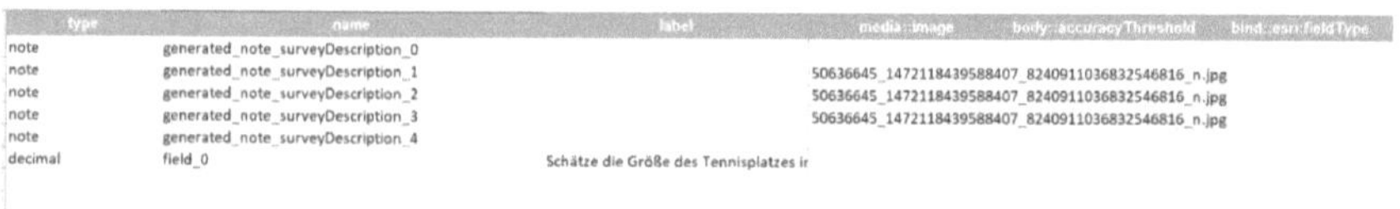

type	name	label	media::image	body::accuracyThreshold	bind::esri:fieldType
note	generated_note_surveyDescription_0				
note	generated_note_surveyDescription_1		50636645_1472118439588407_8240911036832546816_n.jpg		
note	generated_note_surveyDescription_2		50636645_1472118439588407_8240911036832546816_n.jpg		
note	generated_note_surveyDescription_3		50636645_1472118439588407_8240911036832546816_n.jpg		
note	generated_note_surveyDescription_4				
decimal	field_0	Schätze die Größe des Tennisplatzes ir			

Figure 14 Excel structure with two extra images (source: own illustration)

c. Unlocking Web version of Survey123 for geoportal

By looking at the QR code generated links that lead to the mobile app we found out that the survey URL of a Survey from the Geoportal differs from a survey of our private user. They all had some extra information at the end of the URL:?portalUrl=https://geoportal.sbg.ac.at/portal

We tried to add this line also to the webpage of survey123 (https://survey123.arcgis.com/surveys?portalUrl=https://geoportal.sbg.ac.at/portal) and surprisingly we had finally access to the Geoportal version of Survey123 Web. The interface slightly differs with a small black window that gives information about the portal URL. (figure 15) Now we are able to implement all of our original ideas and recreated the surveys in this web version. The surveys are still also available in the connect version as backup copy.

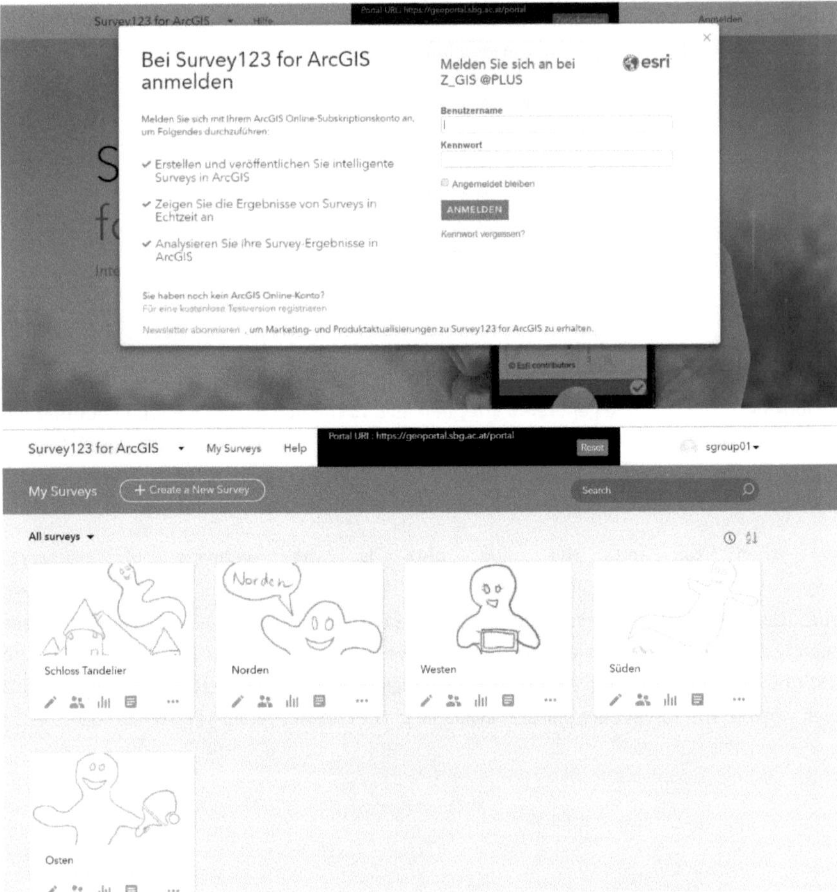

Figure 15 Unlocking Survey123 Web for geoportals (source: Survey123, © ESRI)

8. Conclusion

As proven in this work, GIS systems can be adapted in a way that pupils can interact with GIS applications without prior knowledge. They get in touch with GIS principles in an easy and intuitive way. Concerning software, the introduced apps are not the best to set up a scavenger hunt, as one of them completely fulfils the requirements. However, they are quite handy and straight forward to satisfy their appropriate tasks. Finally, once set up properly, the concept can also be easily adapted to any other region or use case which makes it quite flexible.

Annex

QR-Code Instruction Sheets for 5 groups of pupils (source: own illustration):

Gruppe 1 - Scavenger Hunt Radstadt

Der Geist GISO kann nicht eher ruhen, bevor nicht 5 Aufgaben bewaltigt wurden

- Scannt die QR-Codes um zu den Aufgaben zu gelangen
- offnet dazu die App QR Droid Code Scanner auf den Tablets
- Lest die Aufgabe, folgt den Anweisungen von GISO und schickt eure Antworten ab
- fahrt dann mit der nachsten Aufgabe fort
- Viel Spass :-)

Norden | Osten | Schloss Tandalier | Süden | Westen

Gruppe 2 - Scavenger Hunt Radstadt

Der Geist GISO kann nicht eher ruhen, bevor nicht 5 Aufgaben bewaltigt wurden

- Scannt die QR-Codes um zu den Aufgaben zu gelangen
- offnet dazu die App QR Droid Code Scanner auf den Tablets
- Lest die Aufgabe, folgt den Anweisungen von GISO und schickt eure Antworten ab
- fahrt dann mit der nachsten Aufgabe fort
- Viel Spass :-)

Gruppe 3 - Scavenger Hunt Radstadt

Der Geist GISO kann nicht eher ruhen, bevor nicht 5 Aufgaben bewaltigt wurden

- Scannt die QR-Codes um zu den Aufgaben zu gelangen
- offnet dazu die App QR Droid Code Scanner auf den Tablets
- Lest die Aufgabe, folgt den Anweisungen von GISO und schickt eure Antworten ab
- fahrt dann mit der nachsten Aufgabe fort
- Viel Spass :-)

Gruppe 4 - Scavenger Hunt Radstadt

Der Geist GISO kann nicht eher ruhen, bevor nicht 5 Aufgaben bewaltigt wurden

- Scannt die QR-Codes um zu den Aufgaben zu gelangen
- offnet dazu die App QR Droid Code Scanner auf den Tablets
- Lest die Aufgabe, folgt den Anweisungen von GISO und schickt eure Antworten ab
- fahrt dann mit der nachsten Aufgabe fort
- Viel Spass :-)

Gruppe 5 - Scavenger Hunt Radstadt

Der Geist GISO kann nicht eher ruhen, bevor nicht 5 Aufgaben bewaltigt wurden

- Scannt die QR-Codes um zu den Aufgaben zu gelangen
- offnet dazu die App QR Droid Code Scanner auf den Tablets
- Lest die Aufgabe, folgt den Anweisungen von GISO und schickt eure Antworten ab
- fahrt dann mit der nachsten Aufgabe fort
- Viel Spass :-)